This book belongs to:

The Universe
Galaxies, Solar Systems, Planets, & Space Objects!
Astronomy Words & Coloring Book

Mary Lou Brown
Sandy Mahony

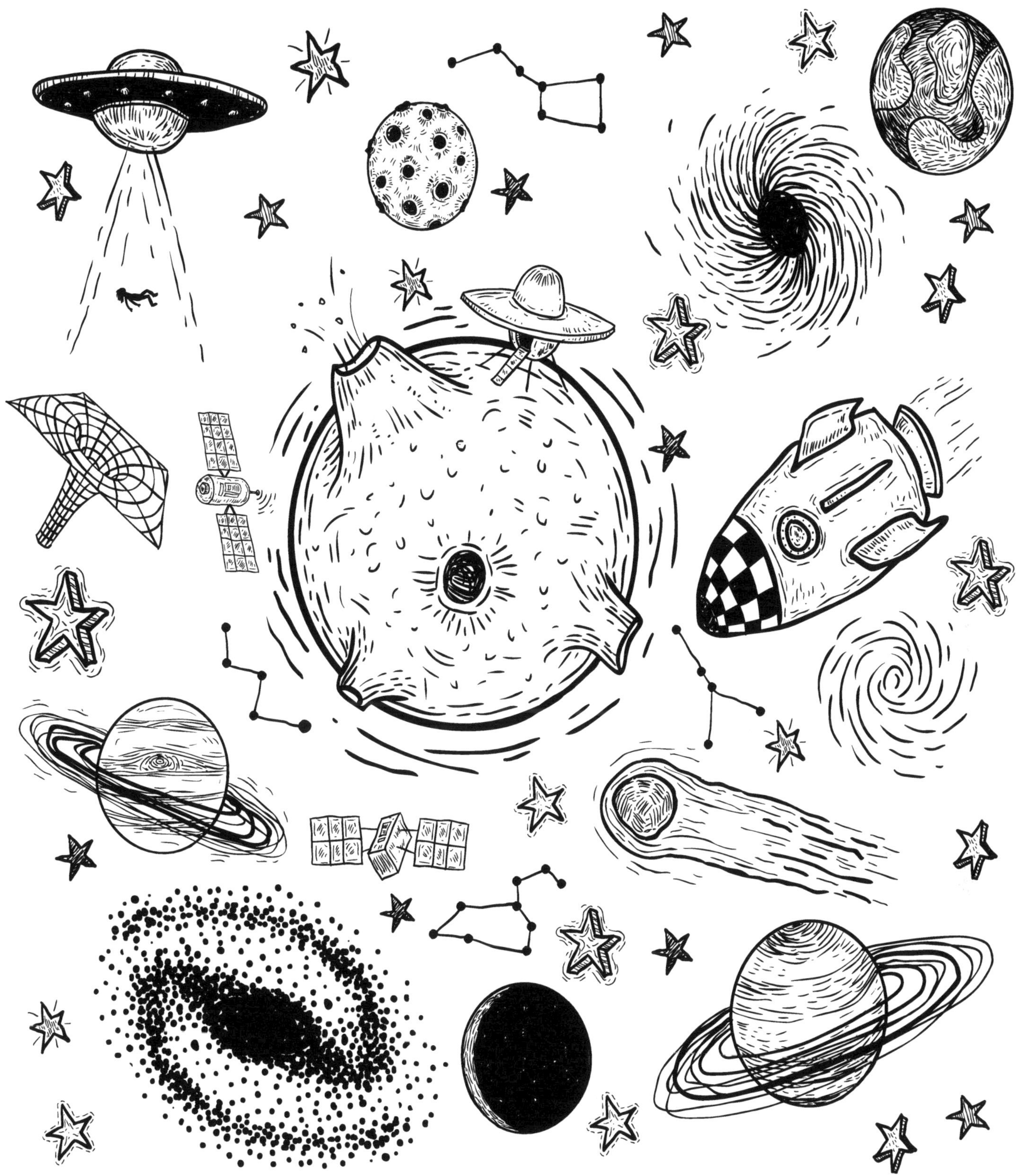

Astronomy

Astronomy is the study of space, stars, and planets.

Space

Space is the region beyond Earth's atmosphere.

Universe

The universe is made up of everything that exists, including Earth and space.

Solar System

The Solar System is made up of the Sun and all the planets, moons, comets and asteroids that move around the Sun.

Sun

The Sun is a star in the center of the Solar System. Earth and other objects orbit around the Sun.

The Sun

- is a gigantic ball of burning gas
- is the biggest object in the Solar System
- has a diameter 110 times larger than Earth
- is located 93 million miles from Earth

Planets

Planets are large round objects that revolve around a star. Our solar system has eight planets.

Mercury

Mercury

- is the closest planet to the Sun in our solar system
- is very hot
- has a surface similar to the Moon

Venus

Venus

- is the second closest planet to the Sun in our solar system
- is similar in size to Earth
- has a surface with many active volcanoes

Earth

Earth

- is third from the Sun and the 5th largest planet in the Solar System
- was formed four and a half billion years ago
- is the only known place in the universe life exists

Mars

Mars

- is the fourth planet from the Sun
- has a barren, rocky surface with a thin atmosphere
- is sometimes called the "Red Planet"

Jupiter

Jupiter

- is the largest planet in our solar system
- has a large number of orbiting moons
- features a Red Spot, a constant storm twice the size of Earth

Saturn

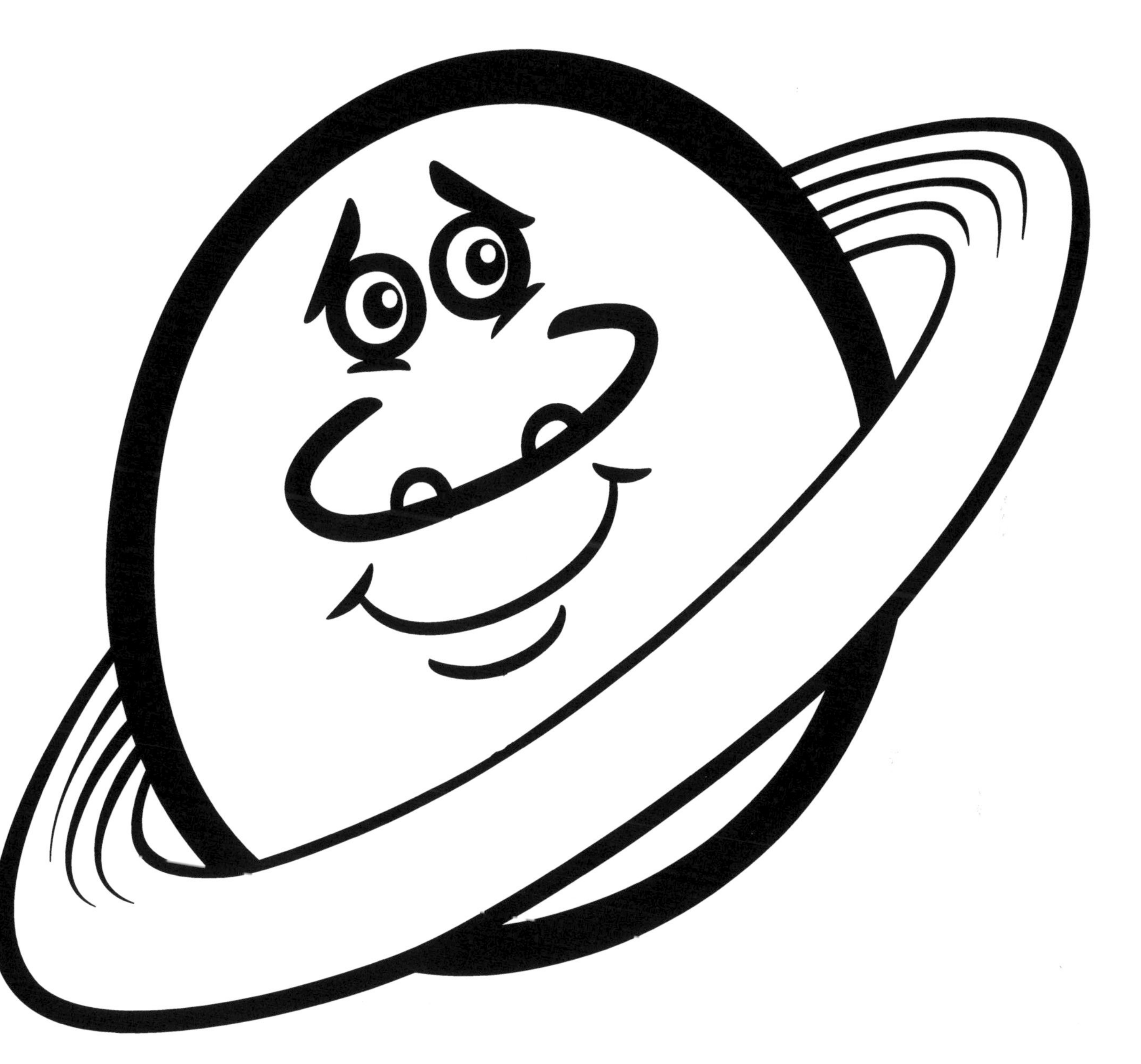

Saturn

- is the second largest planet in our solar system
- is the sixth planet from the Sun
- is known for its system of rings

Uranus

Uranus

- is the third largest planet in the Solar System
- was the first planet discovered by using a telescope
- is the seventh planet from the Sun

Neptune

Neptune

- is the eighth planet from the Sun in our solar system
- is nearly four times the size of Earth
- has strong winds and violent weather

Dwarf Planets

Dwarf planets are objects orbiting the Sun that are not considered planets and are not satellites. Pluto is the largest dwarf planet in our solar system.

Moons

140 moons orbit eight planets in our solars system. The moons don't orbit the Sun. They revolve around the planet they are nearest to.

Earth's Moon

The Moon was probably made 4.5 billion years ago when a large object hit the Earth and blasted out rocks that came together to orbit the Earth. It is airless, waterless, and lifeless.

Moon Landing

The first person to walk on the Moon was American astronaut, Neil Armstrong, who stepped out of his landing craft, the Eagle, on July 21, 1969.

Asteroid

Asteroids are small objects that orbit the Sun. They are made of rock, metal, and sometimes organic compounds (living matter).

Meteor

Sometimes called a "shooting star," meteors are meteoroids that burn up as they pass through the Earth's atmosphere.

Comet

Similar to an asteroid, a comet is a small space object that orbits the Sun. When close enough to the Sun, they display a fuzzy outline and sometimes a tail.

Constellations

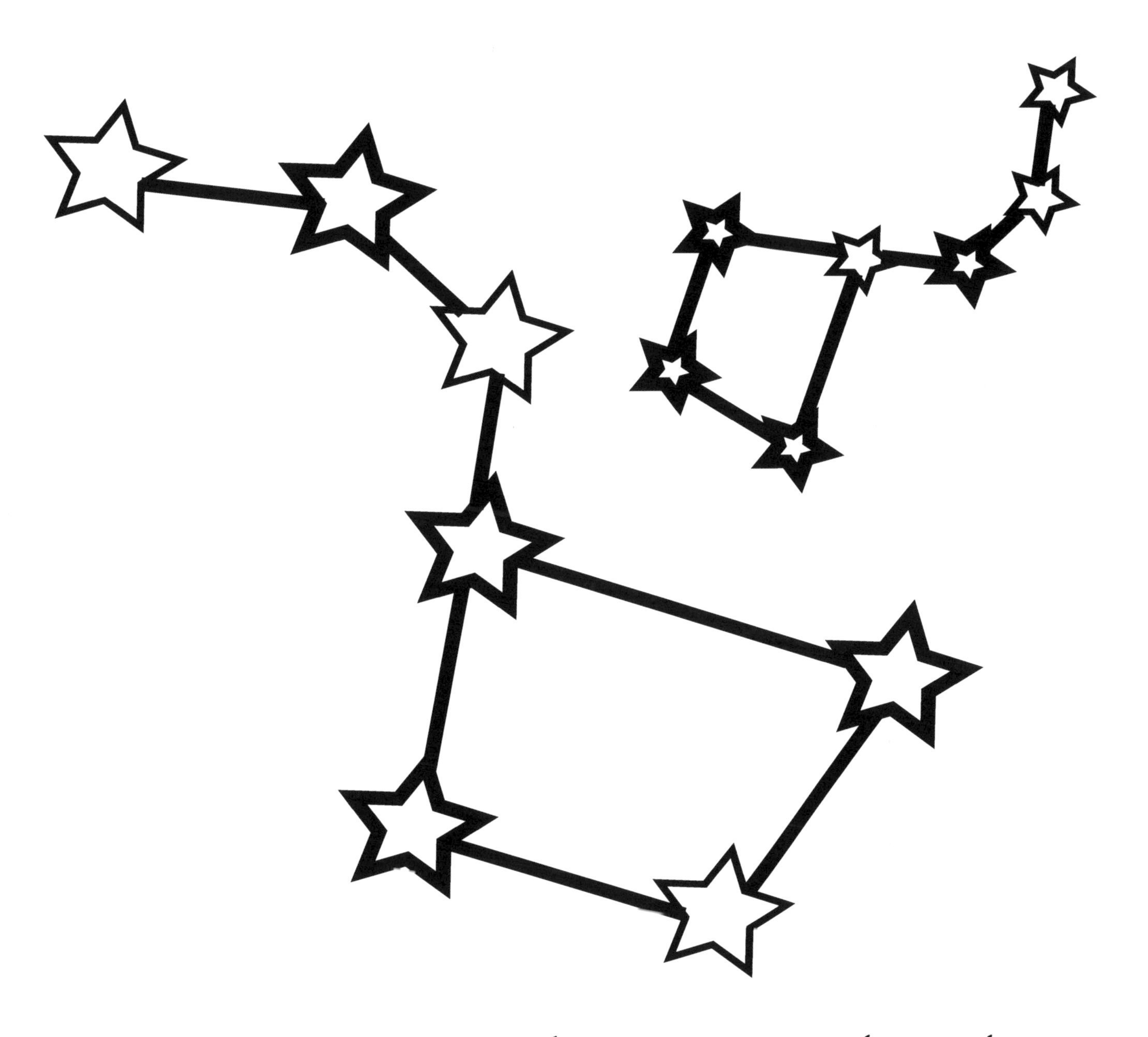

Up to 6,000 stars can be seen at night without a telescope. Constellations are groups of stars in a particular pattern. When people started navigating the oceans, constellations were used to help guide ships from one location to another.

Telescope

In 1609, a scientist named Galileo first peered through his small homemade telescope at the stars. Since then, our knowledge of the universe has grown, and telescopes have grown bigger, better, and are even launched into space.

Astronaut

An astronaut is a person trained to command, pilot, or serve as a crew member of a spacecraft.

Rover

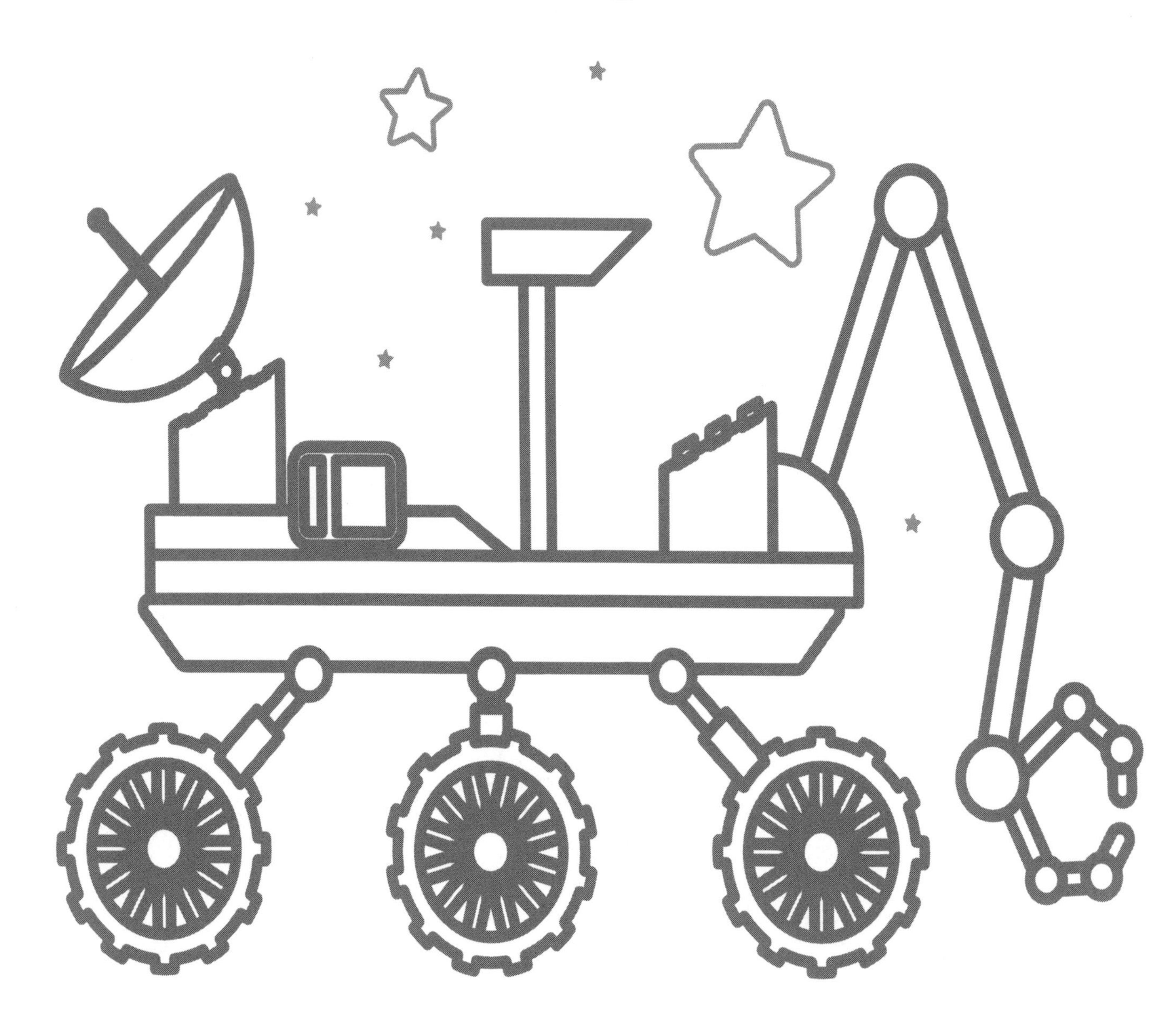

A rover is a space exploration vehicle designed to move across the surface of a planet or other space object. With or without a space crew, rovers are created to land on another planet or moon to find out information and to take samples. They can collect dust, rocks, and even take pictures.

Spaceflight

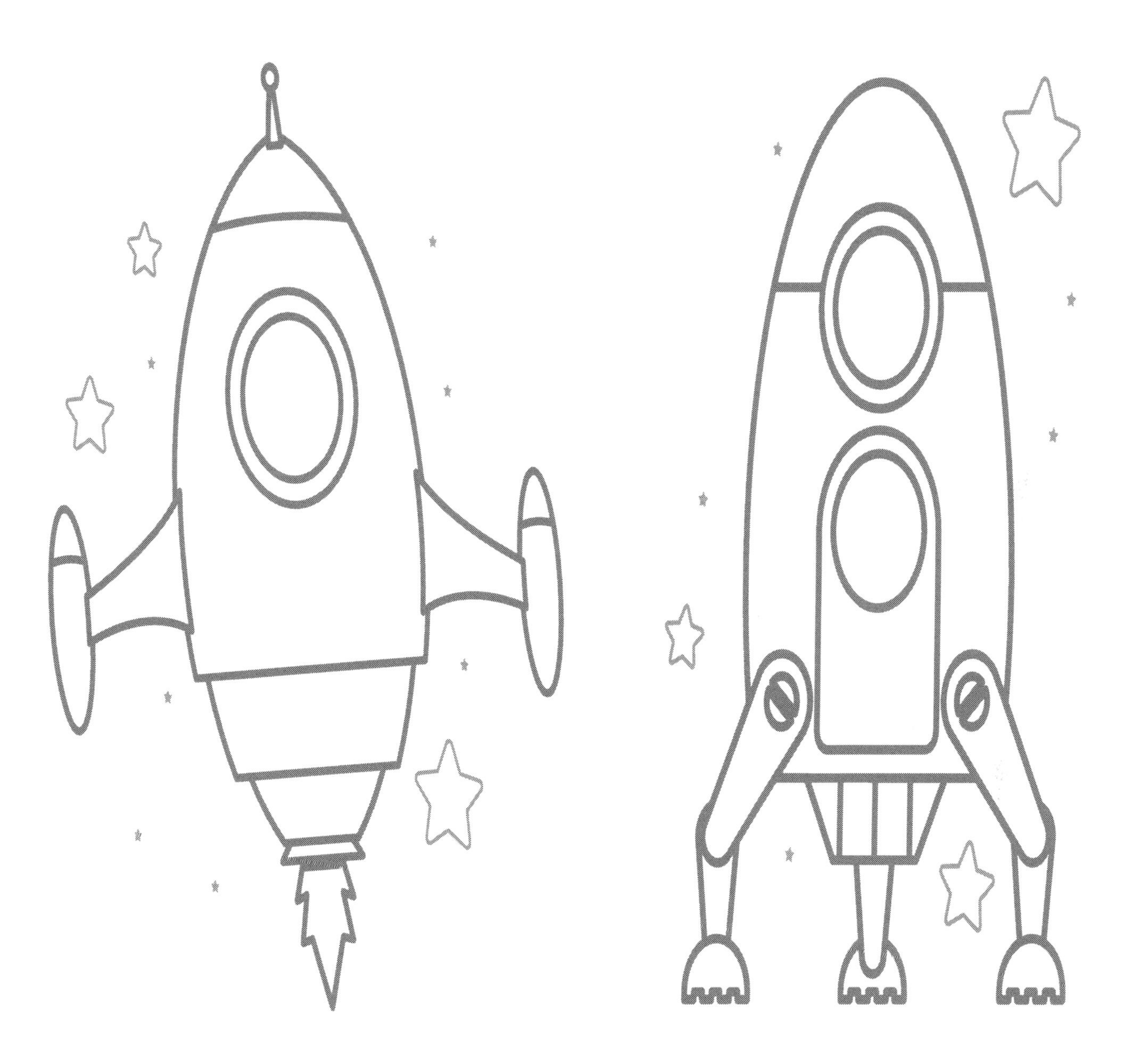

Spaceflight uses space technology to fly spacecraft into and through outer space. Spaceflight is used in space exploration, and also in activities like space tourism and launching satellites.

orbit

An orbit is a path that an object takes in space when it goes around a star, a planet, or a moon. Orbit also means to revolve around another object in space. The Earth orbits the Sun. A satellite is an object in space that revolves around another

satellite

object. The earth is a satellite of the Sun, just like the Moon is a natural satellite of the Earth! Many artificial satellites orbit Earth to send, receive, or bounce back information. They are used for military and civilian observation, weather, research, navigation, and communication.

SPACE

Made in the USA
Monee, IL
23 April 2020

27484705R00020